Self-Concept Selects Your Reality

Also by The Quantum Shifter

The Truth of Reality
The Biggest Misconceptions About Manifestation
Why Isn't It Here Yet?
Manifest In Steps
Identity Shifting
One Decision Creates a New Reality
Self-Concept Selects Your Reality
Manifest Faster
Identity Shifting
I AM PERFECT
Embody a Wealth Identity
Mental Diet Mastery
Stop Spiraling While Manifesting
Emotional Mastery
Tell Yourself This When You're Depressed
Manifesting Money 101
Money From Nothing
You Are The Bank
Money Mindset For Single Moms
Your SP Loves You
Manifest A Celebrity
Have The Best Sex of Your Life
Lose Weight The Mental Way
Manifest Away "ADHD"
The Writer Everyone Reads
You Don't Need 3D Proof
Action Is Optional
The Positive Delusion

Cause and Effect Is a Lie
Objective Reality Is a Myth
Revision
Collapse & Command

SELF-CONCEPT SELECTS YOUR REALITY

Why Reality Changes When Identity Changes

The Quantum Shifter

IDENTITY PRESS

Connect with The Quantum Shifter

Website: www.TheQuantumShifter.com

Facebook: www.Facebook.com/TheQuantumShifter

ISBN: 978-1-972432-19-8

Published by Identity Press

These works were written over many years during the author's ongoing study of identity and reality creation.

They are now being published for the first time under the pen name *The Quantum Shifter*.

YOUR REALITY IS WHAT YOU SAY IT IS.

Contents

Before You Continue

This book will make the most sense if you are open to a simple idea: Your reality is not fixed. It is not happening to you. It is reflecting what you assume is true.

If you believe that reality is solid, unchangeable, and determined by circumstances, parts of this book may feel confusing or difficult to accept. And that's okay. You don't need to force yourself to believe anything. You can simply set this book down and return to it later, if and when it feels right.

But if something in you already senses that reality may not be what it seems—if you've noticed patterns that don't quite make logical sense—or if you've experienced moments where things shifted in ways you couldn't explain— then this book will likely feel familiar.

Not because it's teaching you something new, but because it's putting words to something you already know. Read it slowly. Let it land. And most importantly: Don't try to "do" anything. Just see what you notice.

Who am i being?

Introduction: The Hidden Selector Behind Every Reality

Most people believe they are trying to change their life. They try to change their circumstances. They try to change their thoughts. They try to change how they feel. And when nothing seems to fully shift, they assume they are doing something wrong. But what if the problem isn't effort? What if the problem is that you've been trying to change the reflection instead of the source?

There is a mechanism operating in your life that determines what you experience, what you notice, what repeats, and what feels possible. That mechanism is not your circumstances. It is not time. It is not luck. It is not other people. It is your self-concept.

Self-concept is not just how you feel about yourself. It is not confidence. It is not self-esteem. It is not positive thinking.

Self-concept is the identity you accept as true about who you are. And once that identity is accepted—even subtly—everything in your reality begins to organize around it.

That is why people can: read the same books, use the same techniques, say the same affirmations, ...and get completely different results. Because they are not operating from the same self-concept.

One person says, "I am someone things work out for," and life reflects that.

Another says, "Nothing ever works for me," and life reflects that.

Not because reality is random. But because reality is selective. It reflects the version of you that you assume yourself to be. This is where most people get stuck. They believe they are trying to create something new. More money. A better relationship. A different life. But they are trying to create it from the same self-concept that has already been producing their current experience. So nothing fundamentally changes. They are rearranging details while remaining the same person.

This book is about something much simpler—and much more powerful. It is about understanding that you are not creating your reality piece by piece. You are selecting it. And the selector is your self-concept.

At any given moment, there are countless versions of your life that could be experienced. Different outcomes. Different timelines. Different versions of you. You are not waiting for one of them to appear. You are aligning with one of them based on who you believe you are. That alignment is not something you build over time. It is something you recognize and accept. And once accepted, it becomes natural.

This is why your reality often feels consistent. Not because it is fixed. But because your self-concept has been consistent. You have been seeing yourself in a certain way: as someone who struggles, as someone who is overlooked, as someone who is always figuring things out, or even as someone who is successful in specific areas. And reality has been reflecting that identity with precision.

Not partially. Not occasionally. But continuously. The moment you understand this, something shifts. You stop trying to fix individual problems. You stop chasing outcomes. You stop asking, "How do I make this happen?" And you start asking: "Who am I being in relation to this?"

Because that is the only place change actually occurs.

Self-concept is not something you improve. It is something you decide. It is the story you accept about yourself. And that story determines: what you notice, what you expect, what feels normal, what shows up.

Which means your reality is not responding to effort. It is responding to identity. This is also why change can feel difficult at first. Not because it is complicated. But because you are used to seeing yourself a certain way.

You are used to thinking from that identity. You are used to confirming it through your experiences. So when a different self-concept is introduced, it can feel unfamiliar. Not wrong. Just unfamiliar. And most people mistake that unfamiliarity for resistance. But it's simply the space between who you've been and who you are choosing to be.

This book will show you how to recognize that shift clearly. Not as a process you need to work through. But as a decision you can make.

You'll begin to see that:

your thoughts follow your identity

your reactions reflect your identity

your circumstances mirror your identity

And once that becomes clear, the idea of "trying to manifest" starts to fall away.

Because there is nothing to force. There is nothing to chase. There is only a version of you to step into. And once you do, reality follows. Not slowly. Not eventually. But naturally. Because it always has. The difference is that now, you are aware of what has been selecting it all along.

Your self-concept is not a small part of your life. It is the filter through which your entire life is experienced. And once you understand that, everything becomes simpler. Not because you gain control over every detail. But because you understand the one thing that controls them all.

This is not about becoming a better version of yourself over time. It is about recognizing that the version of you who already has what you want exists now. And choosing to see yourself as that version. Because the moment you do—your reality has no choice but to reflect it.

Chapter 1

What Self-Concept Really Is

When most people hear the term self-concept, they immediately think of things like confidence, self-esteem, or how much they like themselves. They think: "I need to feel better about myself." "I need more confidence." "I need to believe in myself more."

And while those things can be helpful, they are not what actually determines your reality. Because self-concept is not about how good you feel about yourself. It is about who you believe you are. That belief exists whether you are aware of it or not. It exists beneath your thoughts. Beneath your reactions. Beneath your circumstances. And it quietly directs everything.

The Difference Most People Miss

Confidence says:

"I feel good about myself."

Self-esteem says:

"I think I'm worthy."

But self-concept says:

"This is who I am."

That distinction matters more than anything else in this work. Because you can feel confident and still experience patterns you don't want. You can feel worthy and still see the same situations repeating.

Why?

Because your identity—the version of you that you accept as true—has not actually changed. And your reality does not respond to temporary feelings. It responds to identity.

Identity Is the Starting Point

Everything in your life begins with a simple, often unnoticed assumption:

Who am I in this situation?

Am I:

The one who is chosen?

The one who struggles?

The one who figures things out?

The one who is overlooked?

The one who always gets what they want?

You don't usually ask yourself these questions directly. But you answer them constantly through your thoughts, reactions, and expectations. And once that answer is established, your experience begins to match it. Not because reality is testing you. But because reality is reflecting you.

You Are Always Operating From a Self-Concept

There is never a moment where you are not identifying as someone. Even in small, everyday situations, you are carrying a version of yourself.

For example:

If you see yourself as someone who is bad with money, you will:

Think thoughts that reinforce that.

Make decisions that align with that.

Notice evidence that confirms that.

If you see yourself as someone who is always chosen in relationships, you will:

Interpret interactions differently.

Expect certain outcomes.

Feel naturally secure.

In both cases, the behavior follows the identity. Not the other way around. This is where most people get it backwards. They try to change their behavior, hoping it will lead to a new identity. But behavior is a reflection. Identity is the cause.

Why Effort Doesn't Create Lasting Change

This is also why so many people feel like they are "trying" to change their life but nothing sticks.

They try to:

Think more positively, act more confidently, repeat affirmations, stay consistent.

But underneath all of that effort is the same self-concept. So the effort becomes temporary. They might feel different for a moment. They might even see small changes. But eventually, they return to what feels familiar. Because identity always pulls you back to what you believe is true about yourself. Not what you are trying to force.

Self-Concept Is Not Built—It Is Accepted

Another misconception is that self-concept is something you build over time.

That you need to: improve it, strengthen it, develop it. But that's not how it works. Self-concept is something you accept.

You decide—consciously or unconsciously—that a certain version of you is true. And once that decision is made, everything begins to align with it. This is why identity shifts can feel instant. Not because something outside of you changed. But because you stopped identifying as the previous version of yourself. And the moment that happens, your thoughts, reactions, and experiences begin to follow the new identity.

Thoughts Do Not Create Identity—They Reflect It

A lot of teachings focus heavily on thoughts.

"Watch your thoughts."

"Change your thoughts."

"Think better thoughts."

But thoughts are not the root. They are the result.

You don't randomly think the thoughts you think. You think from who you believe you are. If you identify as someone who struggles, your thoughts will naturally reflect struggle.

If you identify as someone who is successful, your thoughts will naturally reflect success. You don't have to force them. They follow automatically.

This is why trying to control every thought can feel exhausting. Because you're trying to override identity instead of changing it.

The Simplicity of It

When you strip everything down, self-concept is simple.

It is:

The version of you that feels normal to be.

Not the version you wish you were.

Not the version you're trying to become.

But the version you quietly accept as:

"This is just who I am."

And whatever that is—no matter how subtle—your reality will reflect it. Not partially. Not occasionally. But consistently.

What This Means Going Forward

Understanding self-concept changes how you approach everything.

Instead of asking:

"How do I make this happen?"

You begin asking:

"Who am I in relation to this?"

Because that answer determines everything that follows.

You don't need to fix every detail of your life. You don't need to control every thought. You don't need to force outcomes.

You only need to recognize:

What version of me am I currently being?

And more importantly:

What version of me am I willing to accept as true instead?

Because the moment that answer changes—your reality begins to change with it. Not because you made it happen. But because you selected a different version of yourself. And reality always reflects the version you choose.

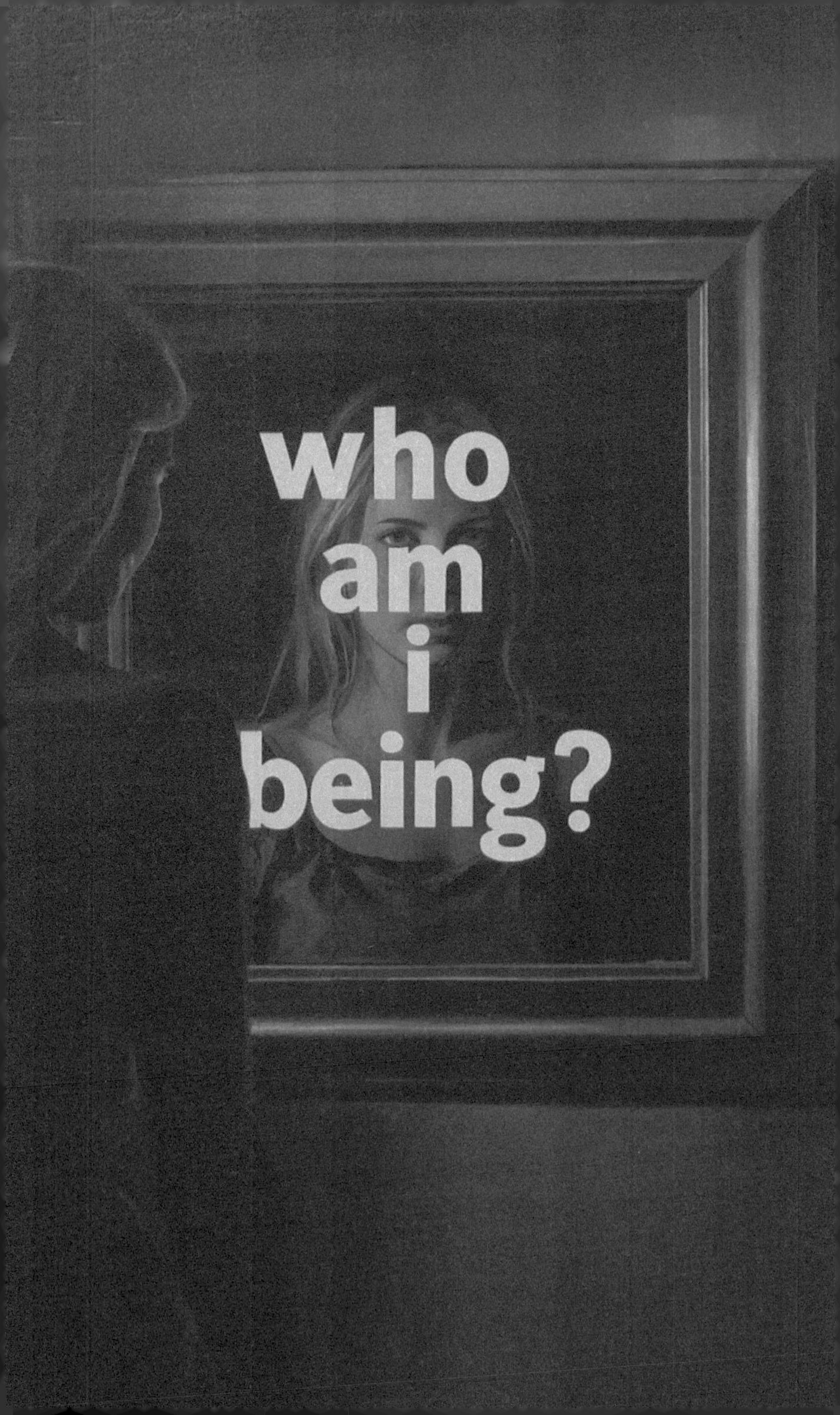
who
am
i
being?

Chapter 2
The Identity Mirror

Reality reflects who you assume you are. Once you understand that self-concept is identity—not confidence, not self-esteem—something important begins to shift. You start to see that your life is not a collection of random events. It is a reflection. Not a perfect, literal mirror where everything is obvious. But a subtle, consistent mirror that reflects back the version of you that you assume yourself to be. This is what we'll call the identity mirror.

Reality Is Not Responding to Effort

Most people believe reality responds to: effort, action, timing, or external circumstances.

They think:

"If I do more, I'll get more."

"If I fix this, things will change."

"If I wait long enough, it will happen."

But if that were true, people who work the hardest would always get the best results. And they don't. Because reality is not responding to what you are doing. It is responding to who you are being. Not in a vague, motivational way. But in a very precise way: **It reflects the identity you assume is true about yourself.**

The Mirror Is Always Active

There is no moment where the identity mirror is turned off. Every experience you have is being filtered through: what you expect, what feels normal, what you believe about yourself.

And then reflected back to you in the form of: circumstances, interactions, outcomes.

This is why patterns exist. Not because life is repeating itself randomly. But because identity is repeating itself consistently.

How the Mirror Works

Let's make this simple.

If you assume: "I am someone who is overlooked,"

you will begin to notice: people not responding to you, being passed over, situations where you feel invisible.

Even if there are moments that don't match that identity, your attention will naturally return to the ones that do. Not because you're trying to focus on them. But because they feel familiar.

Now consider a different identity: "I am someone who is always chosen."

From that identity, you will: interpret interactions differently, expect positive responses, feel secure in situations where you used to feel uncertain. And your reality will begin to reflect that. Not because you forced it. But because you assumed it.

It's Not Just What Happens—It's What You See

One of the most important things to understand about the identity mirror is this: It doesn't just influence what happens. It influences what you notice. Two people can experience the same situation and walk away with completely different conclusions.

Why?

Because they are seeing it through different identities.

One person might say:

"Nothing ever works out for me."

Another might say:

"That actually went pretty well."

Same situation. Different reflection. Because the mirror is not showing "objective reality." It is showing reality as it aligns with identity.

The Subtle Reinforcement Loop

The identity mirror creates a loop. You assume an identity. Reality reflects it. You see the reflection. And that reinforces the identity.

So it feels like: "This is just how life is."

But what you're actually experiencing is: A reflection confirming an assumption.

Over time, this loop becomes so normal that it feels permanent. But it isn't. It is simply consistent. And consistency can be changed.

Why Changing Circumstances Doesn't Work

This is where most people get stuck. They try to change what they see. They try to: fix the situation, control the outcome, improve external conditions. But they do it while holding the same identity.

So even if something changes temporarily, it doesn't last. Because the mirror adjusts back to match the identity. It has to. That's what it does. It reflects.

The Moment Everything Shifts

The shift doesn't happen when circumstances change. The shift happens when identity changes. When you stop assuming: "This is who I am," and begin to accept: "This is who I am now," the mirror has no choice but to adjust.

Not gradually. Not reluctantly. But naturally. Because it is not deciding anything. It is reflecting.

You Don't Have to Force the Reflection

Another important point: You don't need to control how the reflection shows up. You don't need to: manage every detail, plan every step, figure out how it will happen. The mirror doesn't require that. It simply responds to identity.

This is why things often unfold in ways you couldn't have predicted. Because you didn't create the path. You selected the version of yourself—and the path came with it.

The Mirror Is Neutral

The identity mirror is not judging you. It is not rewarding or punishing you. It is not deciding what you deserve. It is neutral. It reflects whatever identity you hold—without question.

Whether that identity is: empowering, limiting, subtle, obvious. It doesn't matter. It reflects it the same way.

Seeing Your Life Differently

When you begin to understand the identity mirror, you stop taking things at face value. **Instead of reacting to what you see, you start asking: "What identity is this reflecting?"**

If you see: struggle, delay, inconsistency, rejection... You don't need to fix those things directly. You look at the identity behind them. Because that is the source.

The Simplicity of Reflection

At its core, the identity mirror is simple: Reality reflects who you assume you are.

Not occasionally. Not selectively. But consistently.

And once you see that clearly, you stop trying to manipulate the reflection. You stop trying to force outcomes. You stop trying to control the external world.

Because you realize:

The only thing that needs to change is the identity being reflected.

And once that changes— the mirror does the rest.

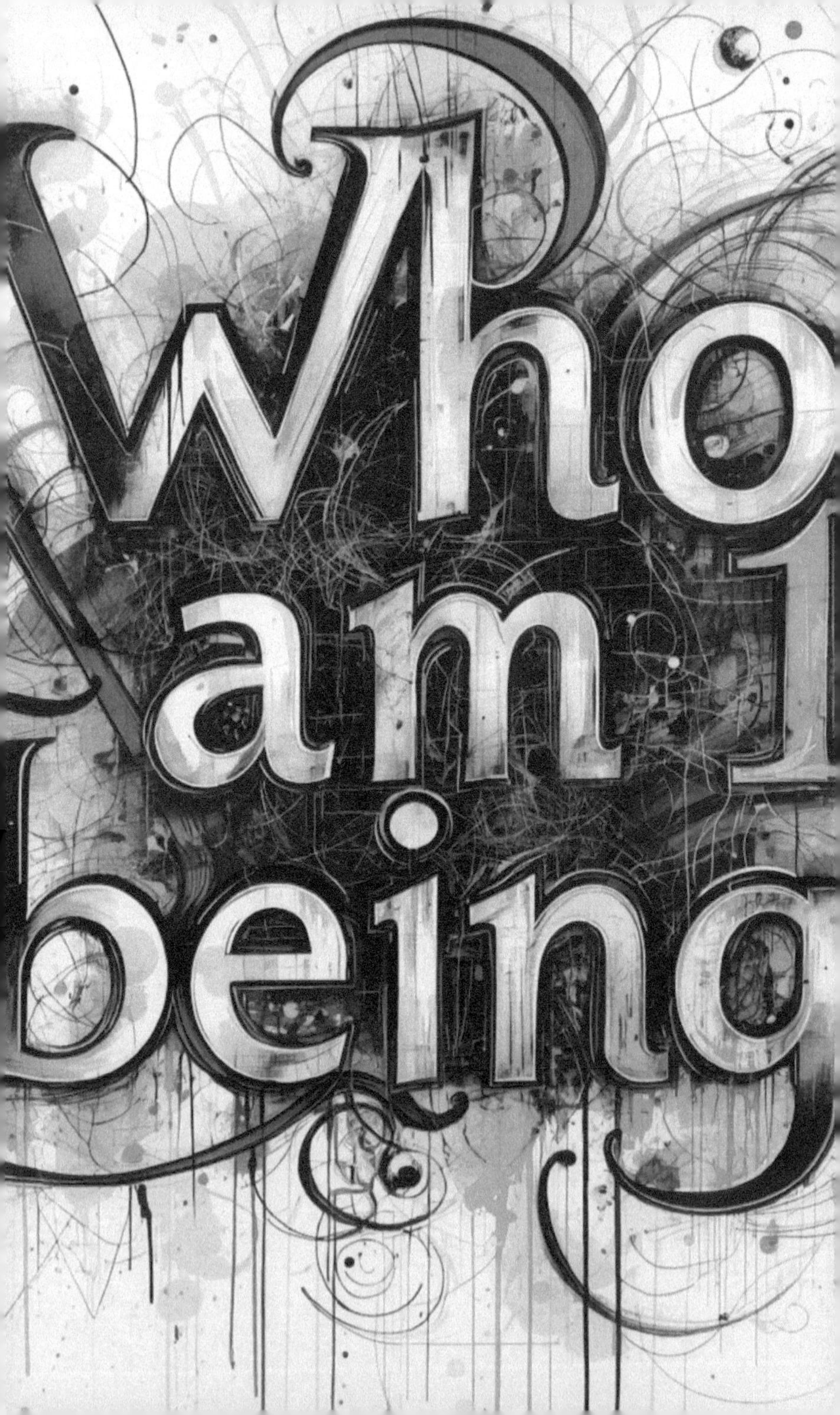
Who am I being

Chapter 3

The Self-Concept Loop

Belief → experience → reinforcement

By now, you've seen that your reality reflects your identity. You've also seen that this reflection is consistent—not random. But there's another layer that makes this feel even more solid, more real, and more difficult to change. That layer is what we'll call the self-concept loop.

This loop explains why your life can feel like it's repeating itself. Why the same types of situations show up again and again. Why even when you try to change, you often find yourself back in the same place. And most importantly: Why your current reality can feel so true.

The Loop in Its Simplest Form

The self-concept loop works like this:

Belief → Experience → Reinforcement

You hold a belief about who you are. That belief shapes your experience. That experience reinforces the belief. And the cycle continues.

Where the Loop Begins

Every loop begins with a belief. Not necessarily something you consciously chose. Often something you absorbed, repeated, or accepted over time.

For example:

"I have to work hard for everything."

"People don't stay."

"Money is inconsistent."

"I'm always figuring things out."

These beliefs don't always sound extreme. In fact, many of them feel normal. That's what makes them powerful. Because when something feels normal, you don't question it. You operate from it.

How Belief Becomes Experience

Once a belief is accepted, it begins shaping your experience in subtle ways. Not by forcing outcomes.

But by influencing: what you expect, how you interpret situations, what feels possible, and how you respond.

For example:

If you believe "money is inconsistent," you may: feel uncertain even when things are going well, hesitate to make certain decisions, or notice fluctuations more than stability.

And those patterns create experiences that reflect inconsistency. Not because the belief caused a specific event. But because it shaped how you moved through your reality.

The Reinforcement Phase

Once the experience occurs, something important happens. You interpret it. And that interpretation reinforces the original belief.

You might say:

"See? This always happens."

Or:

"I knew this wouldn't last."

That moment of recognition is reinforcement. It strengthens the belief. It makes it feel more real. More justified. More permanent.

Why the Loop Feels So Convincing

This is what makes the self-concept loop powerful. It creates the illusion of proof. You don't just believe something. You see it. And because you see it, it feels true.

But what you're actually seeing is:

A reflection of a belief confirming itself. Not objective reality. Not an unchangeable fact. Just a loop continuing.

The Loop Is Self-Sustaining

Once the loop is active, it sustains itself. You don't need to consciously maintain it. It runs automatically. Belief leads to experience. Experience reinforces belief. Belief leads to more experience.

And over time, it becomes:

"This is just how life is."

But what you're really experiencing is:

"This is how my identity is reflecting."

Why Change Can Feel Difficult

When people try to change their life, they usually try to interrupt the loop at the experience level.

They try to: force a different outcome, change circumstances, or control what happens.

But the loop doesn't start there. It starts with belief.

So even if something changes temporarily, the underlying belief remains. And the loop eventually re-establishes itself. This is why people often feel like they've made progress, only to "fall back" later.

They didn't fall back. They returned to the original loop.

Breaking the Loop

To change the loop, you don't fight the experience. You don't argue with the reflection. You don't try to override what you see. You go to the beginning. You change the belief. Not by forcing yourself to believe something new.

But by recognizing that the belief you've been holding is not fixed. It is something you accepted. And anything accepted can be replaced.

The Moment of Interruption

The loop breaks in a very specific moment. The moment you stop saying:

"This is just how it is,"

and start saying:

"This is what I've been assuming."

That shift is subtle, but powerful. Because it moves you from: reacting to reality to understanding what is creating it.

Choosing a New Loop

Once you see the loop clearly, you can choose differently. You can begin with a new belief. Not something you're trying to prove. But something you are willing to accept.

For example:

"Things always work out for me."

"Money is consistent and available."

"I am always chosen."

At first, this may feel unfamiliar. Because it doesn't match the previous loop. But that's expected. You are not continuing the same pattern. You are starting a new one.

Allowing the New Loop to Form

The new belief will begin to shape your experience. You will start to notice: different reactions, different interpretations, and different outcomes.

And as those experiences occur, they will reinforce the new belief. A new loop begins. Not because you forced it. But because you selected it.

The Simplicity of the Loop

At its core, the self-concept loop is simple:

What you believe about yourself becomes what you experience.

What you experience becomes what you believe. And this continues until you choose otherwise.

What This Means for You

You don't need to control every situation. You don't need to analyze every outcome. You don't need to fix every detail. You only need to recognize:

What loop am I currently in?

And:

What loop am I willing to step into instead?

Because the moment that answer changes—the loop changes. And once the loop changes—your reality follows.

Who am I being?

Chapter 4

The Version of You That Exists

You're not creating—you're selecting

Up to this point, you've seen how self-concept shapes experience. You've seen how identity reflects through reality. You've seen how the self-concept loop keeps patterns in place. Now we're going to take one more step deeper. Because there's something even simpler—and more freeing—than everything you've read so far.

You are not creating your reality. You are selecting it.

The Idea That Changes Everything

Most people approach their life as if they are building it piece by piece.

They think:

"I need to create more money."

"I need to create a better relationship."

"I need to create a different future."

And because they see it this way, everything feels like effort.

It feels like:

something to figure out

something to work toward

something that takes time

But what if none of that is actually necessary?

What if the version of you who already has what you want... already exists? Not in a future you need to reach. Not as a possibility you need to build. But as a version of you that is available right now.

Multiple Versions, One Selection

At any given moment, there are countless versions of your life that could be experienced. Different outcomes. Different paths. Different identities.

There is a version of you who: has the money, has the relationship, has the confidence, has the stability.

And there is also a version of you who doesn't.

You are not trying to turn one into the other. You are selecting which version you align with.

Why It Feels Like You're Creating

The reason this often feels like creation is because of time. You've been taught to think in terms of: past → present → future

So it seems like: "I am here now, and I need to get there."

But what you're actually experiencing is not movement through time. It is movement through identity. And identity determines which version of reality becomes visible to you.

Identity Is the Selector

Your self-concept acts like a filter. It determines:

what you notice

what feels possible

what becomes real to you

When you identify as a certain version of yourself, you begin to experience the reality that matches that identity. Not because you created it. But because you aligned with it.

The Illusion of Distance

One of the biggest illusions people experience is the feeling that what they want is far away. That it's something they need to work toward. That it takes time to reach. But that feeling comes from identifying with a version of yourself that does not have it. From that identity, the desired version appears distant.

From the identity that already has it, there is no distance at all. Because from that perspective, it is already normal.

What Selection Actually Looks Like

Selection is not a dramatic moment. It is not something you announce to the world. It is a quiet shift in identity. It is the moment you stop saying:

"This is who I've been,"

and start accepting:

"This is who I am now."

And once that shift happens, everything begins to follow.

You Don't Need to Build the Path

Another important piece of this:

You don't need to figure out how to get from one version to another. Because you're not traveling a path. You're changing identity. And the moment identity changes, the path reorganizes itself. This is why things often happen in unexpected ways. Because you didn't construct the outcome. You selected the version where the outcome exists.

Why This Feels Effortless (When It Clicks)

When you truly understand selection, effort begins to fall away. Not because you stop doing things. But because you stop trying to force outcomes.

You move from:

"How do I make this happen?"

to:

"Who am I being right now?"

And from that identity, actions become natural. Decisions become clear. Experiences begin to align.

The Resistance Most People Feel

When people first hear this, they often feel resistance.

They think:

“That’s too simple.”

Or:

“It can’t be that easy.”

But that reaction is coming from the current identity.

An identity that is used to: effort, struggle, time, process. So the idea of instant selection feels unfamiliar. Not because it’s wrong. But because it doesn’t match what has been normal.

Letting the New Identity Feel Normal

This is where the shift happens. Not by forcing yourself to believe something new. But by allowing a different version of yourself to feel normal. At first, it may feel different. But that’s expected.

You are not continuing the same identity. You are selecting a new one. And as you remain with that identity, it stabilizes. And as it stabilizes, your reality reflects it more consistently.

You Are Always Selecting

Even when you're not aware of it, you are always selecting a version of yourself. Through your thoughts. Through your assumptions. Through what you accept as true.

The difference now is awareness. Instead of selecting unconsciously, you begin selecting intentionally.

The Simplicity of It

At its core, this idea is simple: You are not creating your reality over time. You are selecting the version of reality that matches your identity. And once that identity is accepted, the reflection follows.

What This Means for You

You don't need to build a new life from scratch. You don't need to force change. You don't need to figure out every step.

You only need to recognize:

Which version of me am I identifying as right now?

And:

Which version of me am I willing to select instead?

Because the moment that selection becomes real to you— your reality has no choice but to reflect it. Not because you made it happen. But because you aligned with the version where it already exists.

WHO AM I BEING?

Chapter 5

Why Circumstances Don't Matter

By now, you've seen that your reality reflects your identity. You've seen that patterns repeat because identity repeats. You've seen that you're not creating a new life—you're selecting a version of yourself and experiencing the reality that matches it.

But there's one thing that continues to trip people up more than anything else. Circumstances. What you can see. What has already happened. What appears to be "real" right now.

And it makes sense. Circumstances feel convincing. They feel solid. They feel factual. They feel like proof.

But what you're about to understand is this: Circumstances are not authority. They are evidence of what has already been assumed. And evidence does not control what happens next.

Why Circumstances Feel So Powerful

From a very early point, most people are taught to trust what they can see.

“If it’s happening, it must be true.”

“If it’s there, it must mean something.”

“If it already occurred, it must matter.”

So when circumstances show up—especially unwanted ones—they feel final. They feel like something you have to fix, change, respond to, overcome. And because of that, your attention goes directly to the circumstance. Not the identity behind it.

The Real Role of Circumstances

Circumstances are not causes. They are reflections. They are showing you what has already been accepted as true about who you are. Nothing more. Nothing less.

If you see inconsistency, it reflects an identity that expects inconsistency. If you see delay, it reflects an identity that assumes delay. If you see struggle, it reflects an identity that identifies with struggle.

Not because life is working against you. But because life is reflecting you.

Evidence Only Looks Backward

This is one of the most important shifts you can make. Evidence is always about the past. It is showing you what has already been. It is not telling you what must continue.

But most people treat it as authority over the future. They look at what has happened and think:

"This is what will keep happening."

But that assumption is what keeps the pattern going. Not the circumstance itself.

Why Trying to Fix Circumstances Doesn't Work

When something unwanted appears, the instinct is to fix it. To change the situation directly. To adjust the details. To make the circumstance look different.

But when you do that, you are still operating from the same identity. So even if the situation shifts temporarily, it doesn't last. Because the identity behind it has not changed. And the identity is what determines what shows up.

The Shift Away From Evidence

This is where things begin to change. Not when circumstances improve. But when you stop treating them as authority.

When you stop asking:

"What is happening?"

and start asking:

"What am I assuming that this is reflecting?"

That question takes you out of reaction. And puts you back into understanding.

You Don't Need Proof to Change Identity

One of the biggest misconceptions is that you need evidence before you can believe something new. That you need to see it first. That something has to happen to convince you.

But if that were true, nothing would ever change. Because you would always be waiting for a reflection before allowing the identity to shift. And the reflection only follows the identity.

So instead, the shift happens when you recognize: You don't need proof to select a new version of yourself. You only need willingness.

Acting Without Evidence

This doesn't mean you ignore your life. It doesn't mean you pretend things aren't happening. It means you stop assigning meaning to what you see.

You stop saying:

"This is happening, so it must be true."

And instead you understand:

"This is reflecting what I've been assuming."

From there, you are free to choose differently. Without needing anything to change first.

The Illusion of "But This Is Real"

At this point, it's normal for a thought to come up:

"But this is real."

And it feels that way. Because it's visible. Because it's present. Because it's familiar.

But real does not mean permanent. And real does not mean powerful. It simply means it is currently being experienced.

And what is experienced can change the moment identity changes.

Circumstances Lose Power Immediately

The moment you stop treating circumstances as authority, something shifts. You stop reacting. You stop trying to control. You stop needing them to change first.

And because of that, the identity behind them becomes visible. Once that identity is visible, it can be changed. And once identity changes, the reflection follows.

You Are Not Bound to What You See

This is the freedom in understanding this principle.

You are not bound to:
your current situation

your past experiences
your visible reality

Because none of those things are determining your future. They are reflecting your past identity. And identity is something you can change.

The Simplicity of It

At its core, this is simple:

Circumstances show you what has been assumed. They do not decide what will happen next.

Evidence is not authority. It is feedback. And feedback does not control you.

What This Means Going Forward

You don't need to wait for things to improve.
You don't need to see proof before you believe.
You don't need to fix every detail.

You only need to recognize:

What you are seeing is not in control.

And once you see that clearly—you are free to choose a new identity.

And from that identity—a new reality begins to reflect.

Not because you changed the evidence. But because you stopped letting it define you.

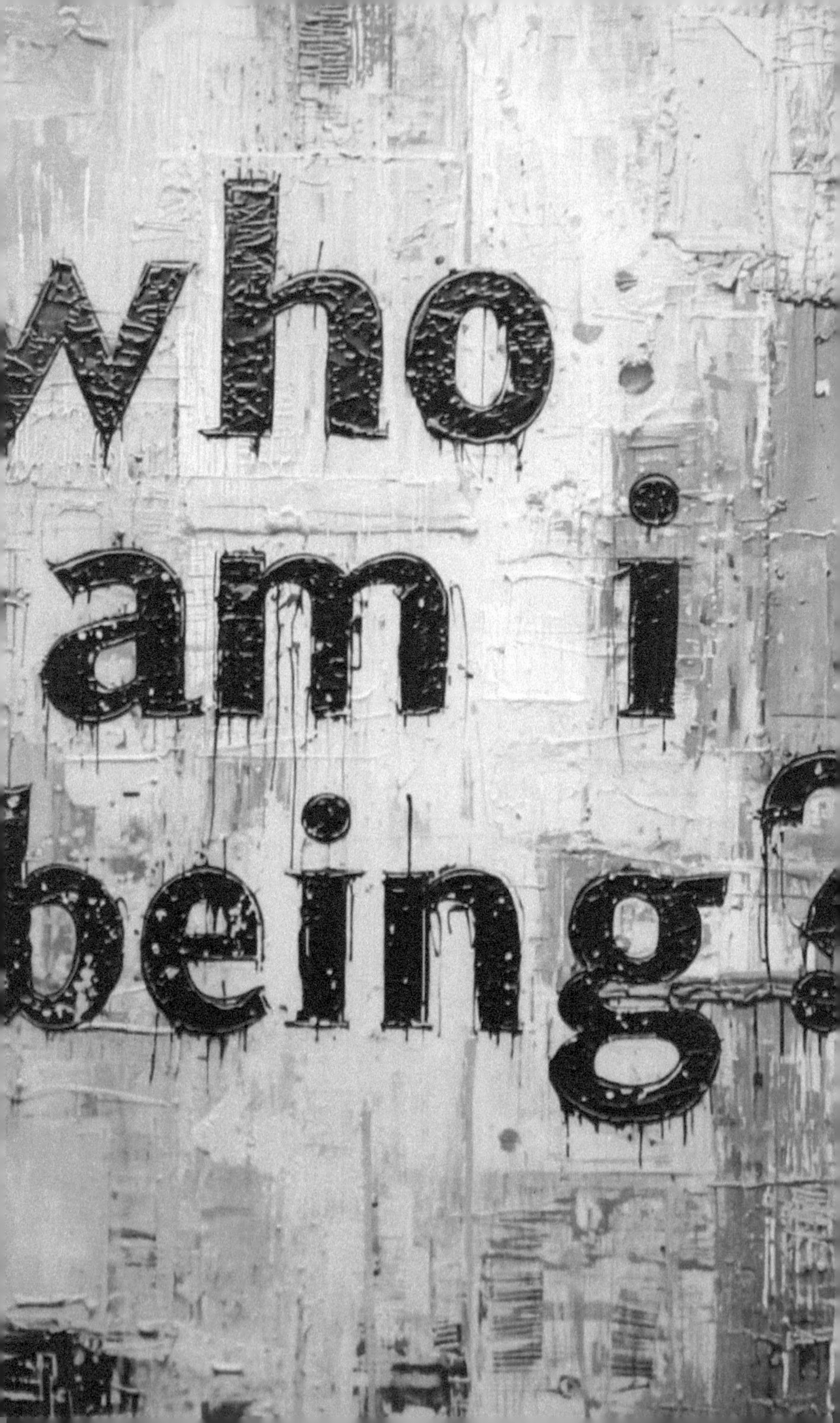
who
am i
being

Chapter 6

The Power of Identity Decisions

One Decision Shifts Identity

By now, you've seen that identity is the source. You've seen that reality reflects who you assume yourself to be. You've seen that circumstances are not in control—they are simply feedback.

So the question becomes:

If identity is what determines everything... how does identity actually change?

Most people assume it changes slowly. That it takes time. That it requires repetition, effort, or consistent practice.

They think:

"I have to work on this."

"I have to become this over time."

"I have to build this identity."

But that assumption comes from the same mindset that sees reality as something you create step by step. And that's not how identity works.

Identity changes in a much simpler way.

It changes through decision.

Identity Is Not Built—It Is Chosen

Think about any identity you currently hold. At some point, whether you realized it or not, it was accepted.

You decided—often unconsciously—that:

"This is who I am."

And once that decision was made, everything began to align with it. Your thoughts followed. Your reactions followed. Your experiences followed.

Not because you built that identity, but because you accepted it.

The Nature of a Decision

A real decision is not something you revisit. It is not something you test. It is not something you question repeatedly.

A real decision feels like:

"This is it."

It has a finality to it. A simplicity. A clarity.

You don't need to convince yourself. You don't need to prove it. You just accept it. And from that point forward, everything adjusts.

Why People Struggle With This

Most people don't actually decide. They attempt. They hope. They try.

They say:

"I'm working on becoming this."

"I'm trying to shift into this."

"I'm getting there."

But all of those statements keep the identity in the future. They keep it separate from who you are now. And as long as it feels separate, it cannot fully reflect, because identity only reflects when it is accepted as current.

The Difference Between Trying and Deciding

Trying sounds like:

"I'm becoming confident."

Deciding sounds like:

"I am confident."

Trying sounds like:

"I'm working on my money situation."

Deciding sounds like:

"I am someone money works for."

Trying keeps the identity out of reach. Deciding makes it immediate.

One Decision Changes Everything

When you truly decide something about who you are, something shifts instantly. Not because the outside world changes immediately, but because your relationship to it does.

Your thoughts begin to align. Your reactions begin to align. Your expectations begin to align.

And because of that, your experience begins to change.

You've Already Done This Before

This is not something new. You've made identity decisions before. You just didn't label them that way.

At some point, you decided:
what you're good at
what you're not good at
what kind of person you are
what you expect from life

And once those decisions were made, they became your baseline. They became normal. And reality reflected them consistently.

Why It Can Feel Unreal at First

When you make a new identity decision, it may feel unfamiliar. Not because it's wrong, but because it doesn't match your previous baseline.

You might think:

"This doesn't feel like me."

But that's exactly the point. You are no longer identifying as the previous version. You are selecting a new one. And anything new will feel different at first.

Letting the Decision Stand

The key is not to keep deciding over and over. The key is to let the decision stand.

Not checking it. Not questioning it. Not looking for proof. Just allowing it to be true.

This is where most people interrupt the shift.

They decide...

...and then they look at their circumstances.

...and then they doubt.

...and then they return to the old identity.

Not because the decision didn't work, but because they didn't let it stabilize.

Stabilization Happens Naturally

When you let a decision stand, something important happens. It begins to feel normal.

Your thoughts begin to reflect it automatically. Your reactions begin to match it. Your expectations adjust. And once it feels normal, reality reflects it consistently.

Not because you forced it, but because you accepted it.

There Is No Process Between Identities

This is another misconception. People believe there is a process between who they are and who they want to be. That there is a path they must follow.

But there is no process. There is only identity. And identity is chosen.

The moment you decide:

"This is who I am,"

you are no longer the previous version, even if your circumstances haven't caught up yet.

The Simplicity of It

At its core, identity decisions are simple. They are not complicated. They are not technical. They are not something you need to master over time.

They are a choice. A clear, simple, final choice.

What This Means for You

You don't need to wait. You don't need to prepare. You don't need to build yourself into a new version.

You only need to decide.

Who are you now?

Not who are you becoming.

Not who are you trying to be.

But who are you now?

Because the moment that answer becomes clear—your identity shifts. And once identity shifts—your reality follows. Not slowly. Not eventually. But naturally.

Because everything in your life has always been reflecting who you decided yourself to be.

And now—you are deciding it on purpose.

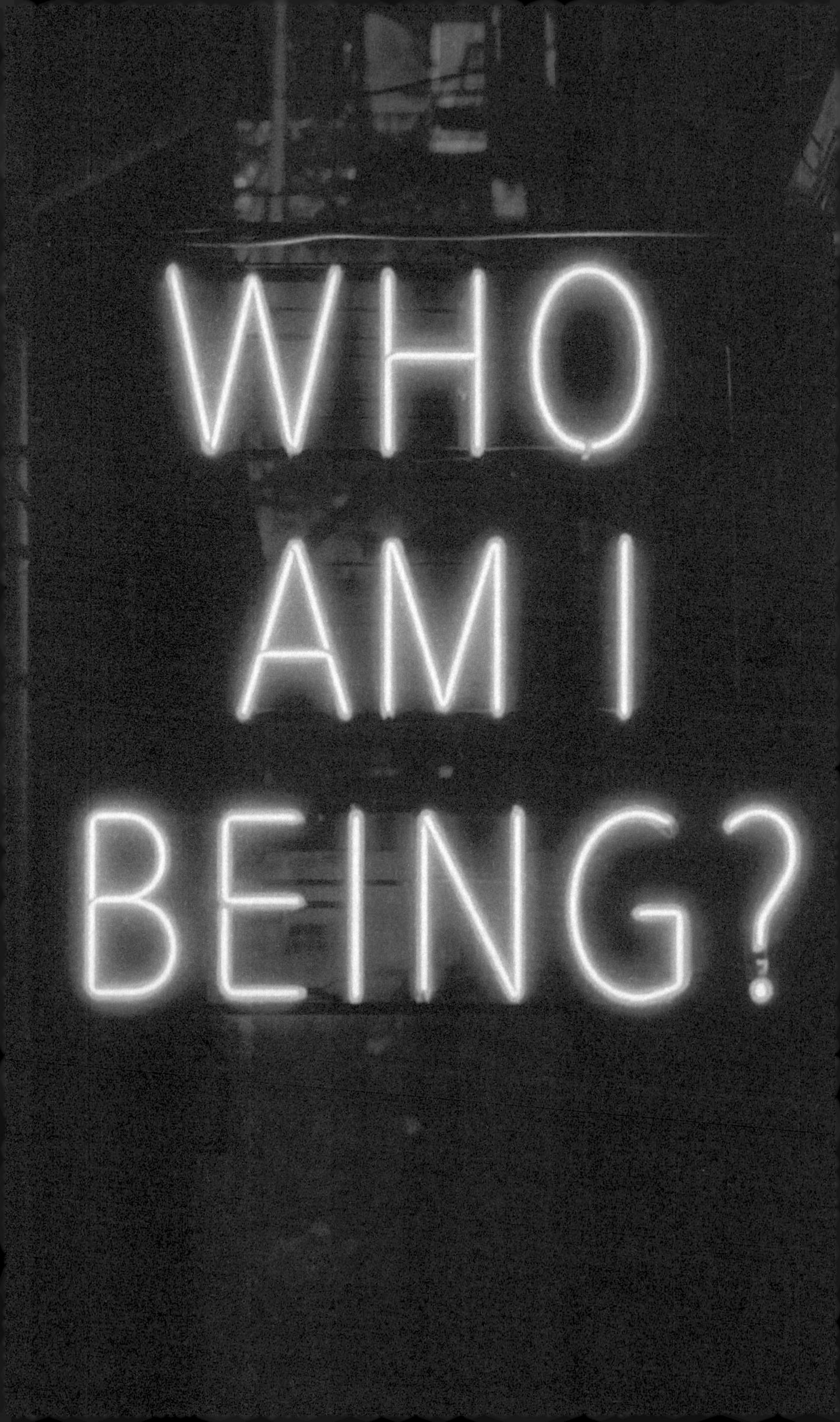
WHO
AM I
BEING?

Chapter 7

Thoughts Follow Identity

You Don't Think Randomly

One of the most common ideas people are taught is this:

"Your thoughts create your reality."

So they try to control their thoughts. They try to think positively, avoid negative thinking, and monitor every idea that comes into their mind.

And while this can feel useful for a short time, it often becomes exhausting. Because thoughts don't actually start the process.

They reflect it.

The Misunderstanding About Thoughts

When people focus only on thoughts, they assume: "If I can just think better, everything will change."

So they watch their mind constantly. They correct themselves. They repeat affirmations. They try to stay "on track."

But underneath all of that effort is the same identity. And identity determines the thoughts that feel natural, not the other way around.

You Don't Think Randomly

Your thoughts are not random. They are not appearing without direction. They are flowing from a source.

That source is your self-concept. **You think from who you believe you are.** Not separately from it.

Identity Generates Thought

Imagine two different identities.

Identity A:

"I am someone who struggles."

Identity B:

"I am someone things work out for."

Each identity will naturally produce different thoughts.

Identity A might think:

"This is probably going to be difficult."

"I hope this doesn't go wrong."

"I always have to figure things out."

Identity B might think:

"This will work out."

"Things usually go smoothly for me."

"I'm not worried about this."

Neither person is forcing those thoughts. They are simply thinking from who they believe they are.

Why Controlling Thoughts Feels Hard

If you try to think from Identity B while still identifying as Identity A, it feels unnatural.

You might say:

"This is going to work out."

But underneath that, there is a feeling of: "I don't really believe that."

So you repeat it. You try harder. You attempt to override the doubt. But the doubt is not the problem. It is a reflection of identity.

You're not fighting thoughts. You're trying to override the source of them.

Thoughts Are a Symptom, Not a Cause

This is where everything becomes simpler. Instead of trying to fix every thought, you recognize:

Thoughts are a symptom of identity.

If the identity changes, the thoughts follow. Automatically. Without effort.

Natural Thinking vs. Forced Thinking

There is a clear difference between natural thoughts and forced thoughts.

Forced thoughts feel like:
effort
repetition
correction
control

Natural thoughts feel like:
ease

normalcy
familiarity
consistency

When you shift identity, your thoughts move from forced to natural. You don't need to monitor them. They align on their own.

The Thought Pattern Shift

When identity changes, your internal dialogue begins to change as well.

You may notice:
you react differently
you assume different outcomes
you feel less resistance

Not because you're trying to think differently, but because you are now thinking from a different version of yourself.

Why Some Thoughts Keep Repeating

If certain thoughts keep appearing, it's not because you haven't controlled them well enough. It's because they are tied to an identity you are still holding.

For example:

If you repeatedly think:

“This might not work,”

it’s likely connected to an identity that expects inconsistency or difficulty.

Trying to silence that thought without changing the identity will only create tension. Because the identity is still active.

Changing Identity Changes Thought Flow

The moment you accept a different identity, something shifts.

Your thoughts begin to:
align with the new identity
support the new identity
reinforce the new identity

And they do this without effort. This is why identity is the focus. Not thought control.

You Don’t Need to Fix Every Thought

Another important realization:

You don’t need to correct every thought. You don’t need to catch every negative idea. You don’t need to be perfect.

Because thoughts are not the root. They are reflections. When the root changes, the reflections change.

What to Do Instead

Instead of asking:

"How do I think better?"

Ask:

"Who am I being right now?"

Because that answer determines what thoughts will follow.

If you identify as someone who is secure, your thoughts will reflect security.

If you identify as someone who is successful, your thoughts will reflect success.

You don't need to force that. It happens naturally.

The Illusion of Mental Effort

Trying to control every thought creates the illusion that reality depends on constant effort. That you need to manage your mind perfectly.

But that's not how this works.

Reality reflects identity. And identity shapes thought. So once identity is clear, the need for effort drops away.

The Simplicity of It

At its core, this idea is simple:

You don't think randomly.

You think from who you believe you are.

And once that belief changes, your thoughts change with it.

What This Means Going Forward

You don't need to fight your mind. You don't need to control every thought. You don't need to monitor yourself constantly.

You only need to recognize:

What identity am I thinking from?

And:

What identity am I willing to think from instead?

Because the moment that identity shifts—your thoughts follow. And when your thoughts follow—your reality continues to reflect that identity with ease. Not because you controlled it. But because you aligned with it.

WHO
AM I
BEING?

Chapter 8

The Story You Carry About Yourself

Narratives Drive Reality

There is something running quietly in the background of your life. You don't always notice it. You don't always question it. But it is always there.

It is the story you carry about yourself.

The Constant Inner Narrative

Everyone is telling a story. Not out loud, necessarily. But internally.

Moment by moment, you are narrating:
what is happening
what it means
what it says about you

You might say:

"This always happens to me."

"I'm just not that kind of person."

"I have to figure everything out."

"Things take time for me."

These statements can feel small. They can feel casual. But they are not insignificant.

They are defining your identity.

And your identity is selecting your reality.

The Story Feels Like Truth

The reason this story is so powerful is because it feels true.

It feels like:
an observation
a conclusion
a fact

But what you are experiencing is not truth.

It is repetition.

You have told the same story enough times that it has become familiar. And what is familiar feels real.

How the Story Forms

Your story didn't appear all at once. It formed over time.

Through:
past experiences
interpretations
repeated thoughts
reinforced beliefs

Something happened. You interpreted it. You told yourself what it meant. And then you repeated that meaning.

Eventually, it became:

"This is just how it is."

But that conclusion was never fixed. It was accepted.

The Story Shapes Everything

Once the story is in place, it begins to influence everything. Not in obvious ways. But in subtle, consistent ways.

It shapes:
what you expect
how you respond

what you notice
what feels possible

And because of that, your experience begins to match the story. Not because the story is true. But because it is being told.

You Don't Experience Reality Directly

This is an important realization. You are not experiencing reality directly. **You are experiencing reality through the story you are telling about it.**

Two people can go through the same situation and walk away with completely different experiences. Because they told different stories about it.

One might say:

"That didn't go well."

Another might say:

"That actually worked out."

Same situation. Different narrative. Different reality.

The Story Reinforces Itself

Just like the self-concept loop, the story reinforces itself.

You tell a story. Reality reflects it. You see the reflection. And that strengthens the story.

So it begins to feel like:

"This keeps happening because it's true."

But it keeps happening because it keeps being told.

Why Changing Circumstances Doesn't Change the Story

Many people try to change their life by changing their circumstances.

They think:

"If I fix this, I'll feel different."

But if the story remains the same, the experience returns to match it. Because the story is what defines the identity. And identity is what selects reality.

The Story Is Always About You

Even when your story seems to be about other people or situations, it is always about you.

For example:

“They never choose me”

is really

“I am someone who is not chosen”

“Money is hard to keep”

is really

“I am someone who experiences unstable money”

The story may appear external. But it is always rooted in identity.

You Can Hear the Story If You Listen

If you pause and pay attention, you can hear your story. In your thoughts. In your reactions. In the way you describe your life.

It shows up in phrases like:

“I always...”

"I never..."

"That's just how it is for me..."

"I'm the kind of person who..."

These are not random statements. They are identity declarations.

The Story Is Not Fixed

This is where everything begins to shift. The story you carry about yourself is not permanent. It is not something you are bound to. It is something you have been telling.

And anything that can be told can be changed.

Changing the Story

Changing your story does not mean denying your experience. It means changing what your experience means about you.

Instead of saying:

"This always happens to me,"

you can say:

"That was a version I experienced, but it's not who I am."

Instead of:

"I have to figure everything out,"

you can say:

"Things work out for me naturally."

You are not rewriting the past. **You are rewriting the identity you are carrying forward.**

The New Story Feels Different at First

When you shift your story, it may feel unfamiliar. Because it doesn't match what you've been repeating. But unfamiliar does not mean untrue. It means new. And new is where change begins.

Letting the Story Stabilize

The key is not to keep switching stories. It is to choose one and let it stabilize. Let it become normal. Let it become the lens through which you see your life. Because once it feels normal, reality reflects it consistently.

The Simplicity of It

At its core, this is simple:

The story you carry about yourself becomes the reality you experience.

Not because the story is powerful on its own. But because it defines your identity. And identity selects reality.

What This Means for You

You don't need to fix every situation. You don't need to control every outcome. You only need to become aware of the story you are telling.

And ask:

Is this the story I want to continue living?

Because if it isn't—you are free to choose a new one.

And the moment you do—your reality begins to reflect that story instead. Not because you forced it. But because you told a different story about who you are. And reality always listens.

WHO
AM I
BEING?

Chapter 9

Declaring a New Self-Concept

Your word = identity selection

Up to this point, you've seen how identity shapes experience. You've seen how the story you carry becomes your reality.

You've seen how thoughts follow identity, and how circumstances reflect what has already been assumed.

Now we come to something very direct. Very simple. And very powerful.

Your word selects your identity.

Not over time. Not gradually. But in the moment you accept it.

What a Declaration Really Is

A declaration is not just something you say. It is something you decide is true.

There is a difference between:

repeating words, hoping something becomes real, trying to convince yourself, and declaring something as fact.

A declaration feels like:

“This is how it is.”

There is no negotiation in it. No waiting. No checking.

It is a decision expressed through words.

Words Carry Identity

Every time you say something about yourself, you are selecting an identity. Even if it feels casual. Even if it feels automatic.

Statements like:

“I always have to figure things out.”

“Nothing works out for me.”

“I'm bad with money.”

“I'm just not that kind of person.”

These are not observations.

They are declarations. And once declared, they become the identity you operate from.

Why Your Word Matters

Your word is not just communication. It is selection.

It tells your mind:

"This is who I am."

And once that is established, everything begins to align with it.

Your thoughts follow. Your expectations follow. Your experiences follow.

Not because your words are magically creating something. But because they are defining identity.

The Moment of Selection

When you declare something with clarity, something shifts immediately. Not necessarily in your circumstances right away. But in your identity. And identity is what determines everything that follows.

For example:

If you say:

"I am someone things always work out for,"

and you mean it as a decision—not a hope— you have just selected a new identity.

From that point forward, your thoughts begin to align with that identity.

Your reactions shift. Your expectations adjust. And your reality begins to reflect it.

The Difference Between Saying and Declaring

This is where most people get stuck. They say things. But they don't declare them.

They repeat:

"I am confident."

"I am successful."

"I am abundant."

But underneath, they are still questioning. Still checking. Still looking for proof. So the words don't land as identity.

They remain separate from who they believe they are.

A Real Declaration Feels Different

A real declaration doesn't feel like effort. It feels like a decision. It feels final. Not loud or dramatic. Just clear.

You don't feel the need to repeat it over and over. You don't feel the need to prove it. You simply accept it.

You Are Already Declaring Constantly

Even if you're not aware of it, you are always declaring something. Through your thoughts. Through your reactions. Through your internal dialogue.

When you think:

"This is going to be difficult,"

that is a declaration.

When you think:

"This probably won't work,"

that is a declaration.

And those declarations select identity just as effectively as intentional ones.

Choosing Your Word Intentionally

The shift happens when you become aware of this.

When you recognize:

"I am already declaring who I am."

And you begin to do it intentionally.

<u>Instead of repeating old statements automatically, you choose:</u>

"I am someone things work out for."

"I am always chosen."

"Money flows to me easily."

"Everything aligns in my favor."

Not as wishes. Not as affirmations you're trying to believe. But as identity selections.

You Don't Need to Argue With the Old

Another important point: You don't need to fight your old declarations. You don't need to correct every thought. You don't need to analyze what you've been saying. You simply choose something new.

Because identity is not maintained by argument. It is maintained by acceptance.

The moment you accept a new declaration as true, the previous one loses its hold.

Letting the Declaration Stabilize

Once you declare a new identity, the key is to let it stabilize. Not checking for immediate results. Not looking at circumstances for confirmation. Not questioning it. Just allowing it to be true.

As it stabilizes, your internal experience shifts. Your thoughts begin to match it. Your reactions align. And your reality follows.

The Simplicity of It

At its core, this is simple:

What you declare about yourself becomes the identity you live from.

Not because words have power on their own. But because they define what you accept as true.

What This Means for You

You don't need to search for the right method. You don't need to wait for the right moment. You don't need to build a new identity slowly.

You can choose it. Through your word. Through a simple, clear declaration.

Who are you now?

Not in the future. Not in theory. But now. Because the moment that answer becomes clear—your identity shifts.

And when your identity shifts— your reality begins to reflect it.

Not because you forced it. But because you declared it. And your word is what selected it.

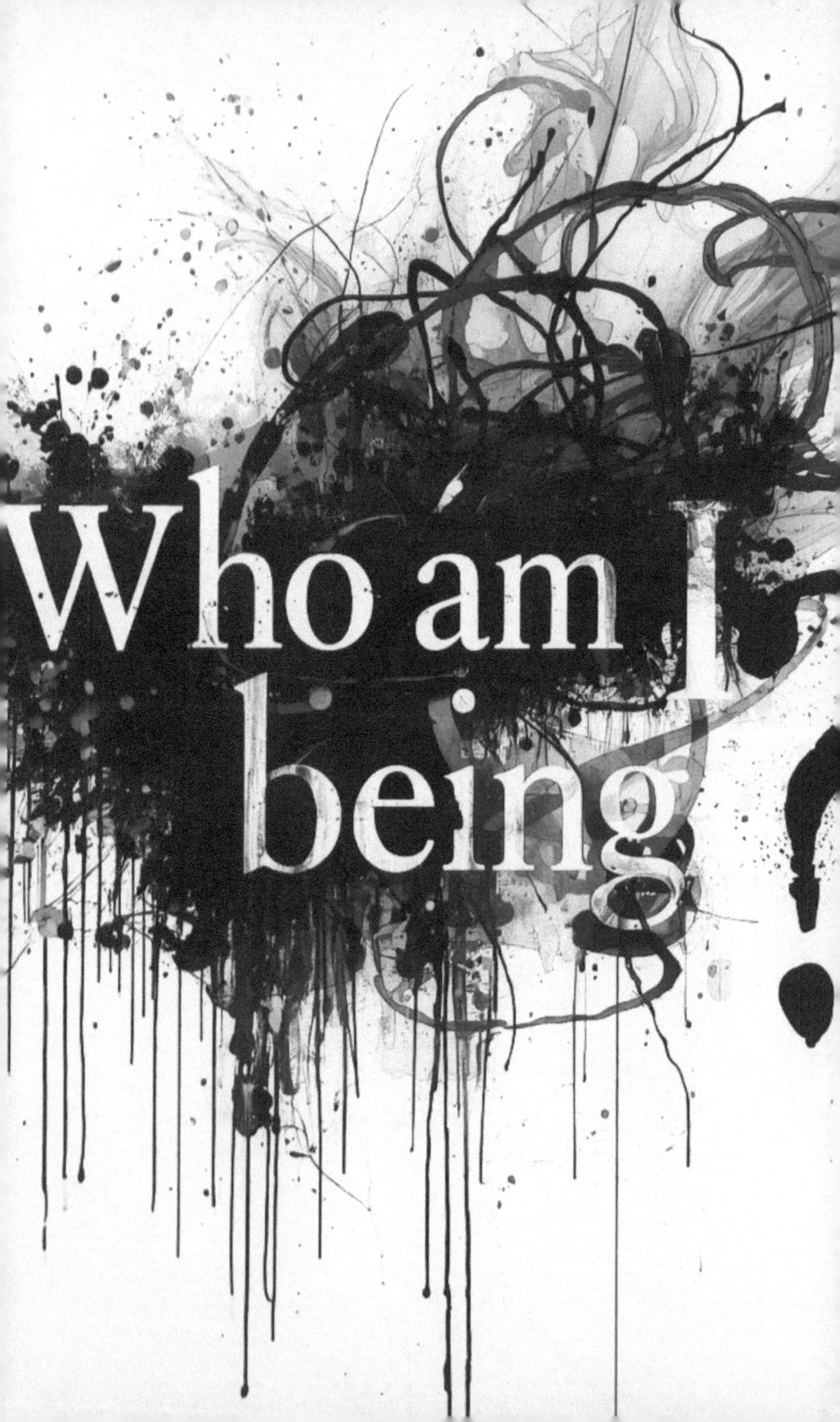
Who am I
being?

Chapter 10

Living as the Version You Selected

Once you understand that identity is selected—and that your word defines that identity—the next question naturally becomes:

What does it actually look like to live as this version of me?

Not in theory. Not in imagination. But in everyday life. Because this is where many people feel uncertain.

They think:

"Do I need to act differently?"

"Do I need to change my behavior?"

"Do I need to force myself to do things I wouldn't normally do?"

And this is where it's important to be clear.

Living as the version you selected is not about forcing behavior. It is about allowing behavior to follow identity naturally.

Behavior Is a Reflection, Not a Strategy

Most people try to change their behavior first.

They think:

"If I act like the version of me who has what I want, then I'll become that person."

So they push themselves.

They try to:

act more confident, make different choices, behave in ways that don't feel natural yet

But when behavior is forced, it creates tension. Because it's not coming from identity. It's coming from effort. And effort cannot sustain what identity has not accepted.

Identity Leads, Behavior Follows

When identity shifts, behavior follows automatically.

Not all at once. Not dramatically. But consistently.

For example:

If you genuinely see yourself as someone who is calm and secure, you won't need to force yourself to react differently.

Your reactions will change naturally.

If you see yourself as someone who always has money available, your decisions around money will begin to shift without effort.

You won't need to think about it constantly. It will feel normal.

The Subtle Nature of the Shift

One of the most important things to understand is that this shift is often subtle.

It's not always:

"I am completely different overnight."

It's more like:

you think slightly differently

you respond slightly differently

things that used to bother you don't feel the same

choices feel easier

And over time, those small shifts create noticeable changes in your experience.

You Don't Need to Perform the Identity

A common mistake is trying to perform the identity.

To act it out. To exaggerate it. To prove it. But real identity doesn't need to be performed. It feels normal.

It feels like:

"This is just who I am."

There is no need to convince anyone. There is no need to act differently for the sake of it. Because behavior that comes from identity is effortless.

The Difference Between Forcing and Allowing

Forcing behavior feels like:

"I should do this."

"I need to act this way."

"I have to change."

Allowing behavior feels like:

"This is what makes sense now."

"This feels natural."

"This is just what I would do."

That difference matters. Because only one of those is sustainable.

Your Decisions Begin to Shift

When you live as a new version of yourself, your decisions begin to reflect it. Not because you are analyzing everything. But because your baseline has changed.

What used to feel normal may no longer feel aligned. What used to feel difficult may now feel simple. And what once required effort may now feel obvious.

You Stop Overthinking

Another shift that often happens:

You stop overthinking your actions. Because overthinking comes from uncertainty. When identity is clear, there is less uncertainty. You don't need to question every move. You don't need to analyze every situation.

You simply respond as the version of yourself you have accepted.

The World Responds Differently

As your behavior shifts—even subtly—your environment begins to respond differently.

People interact with you differently. Opportunities show up differently. Situations unfold in ways that match your new identity.

Not because you controlled the outside world. But because your identity changed how you move through it.

There Is No Checklist

Living as the version you selected is not about following steps. There is no checklist. There is no perfect way to do it. Because it is not a method. It is an identity. And identity expresses itself naturally.

Trusting the Process

At this point, it's important to trust the process. Not as something external. But as something that is already happening. Once identity shifts, everything begins to adjust. Even if you don't notice it immediately. Even if circumstances haven't fully caught up. The shift is already in motion.

You Don't Need to Get It Perfect

Another important point:

You don't need to do this perfectly. You don't need to monitor yourself constantly. You don't need to correct every action. Because perfection is not required.

Consistency in identity is what matters. And consistency comes from acceptance.

The Simplicity of Living as the New Version

At its core, this is simple:

You are not trying to act like a new version of yourself.

You are allowing yourself to be that version.

And from that, behavior naturally aligns.

What This Means for You

You don't need to force change. You don't need to push yourself into different actions. You only need to remain clear on who you are. Because once that is clear, everything else follows.

Your thoughts align. Your behavior aligns. Your experiences align. Not because you made them. But because you selected the version of yourself that naturally lives that way.

And once that selection is real— living as that version becomes effortless. Because it's no longer something you're trying to do.

It's simply who you are.

Who am I being?

Chapter 11

Stabilizing Your New Reality

By now, you've seen how identity selects reality. You've seen how one decision can shift who you are. You've seen how thoughts follow identity and behavior aligns naturally.

So the question becomes:

If it's this simple, why do people feel like they "lose it"?

Why does it sometimes feel like you've shifted... and then something pulls you back?

Why do old thoughts, old reactions, or old circumstances seem to reappear?

This is where understanding stabilization becomes important.

What "Losing It" Really Is

When people say they've "lost" their new identity, what they're actually experiencing is this:

They returned their attention to the old one. Not because the new identity didn't work. Not because something went wrong. But because the old identity still feels familiar. And familiarity has a strong pull.

The Pull of the Old Identity

The identity you've lived in for a long time has momentum.

It has:

familiar thoughts, familiar reactions, familiar expectations

So when something happens that resembles the old pattern, it can trigger a moment of recognition:

"This feels like before."

And if that moment is accepted, the mind begins to reconnect with the old identity. Not fully. But enough to feel like a shift back.

The Role of Attention

Stabilization is not about effort. It is about attention. Where your attention goes, identity is reinforced.

If your attention goes to: old interpretations, old meanings, old assumptions - you begin to re-activate the previous identity.

If your attention remains with the new identity, it stabilizes.

You Are Not Starting Over

This is an important point. You are never starting over. Even if old thoughts appear. Even if old reactions happen. Even if something looks like the past.

None of that erases the identity you selected. It only matters what you continue to identify with.

The Moment of Choice

Every time something happens, there is a subtle moment of choice.

Not a dramatic decision. Just a quiet recognition:

"Am I going to interpret this from my old identity, or my new one?"

That moment determines whether the new identity stabilizes—or the old one is reinforced.

Old Patterns Can Appear Without Meaning

One of the biggest misunderstandings is thinking that old patterns mean something.

That if something looks like the past, it means:

"It didn't work."

But that's not true.

Old patterns can appear briefly without being re-established. They only gain power if you assign meaning to them.

If you say:

"This is happening again," you reconnect to the old identity.

If you say:

"This is irrelevant to who I am now," it loses its hold.

Stability Comes From Non-Reaction

Stabilizing a new identity does not require constant effort. It requires less reaction.

Not reacting to: old thoughts, old assumptions, old appearances.

Not by suppressing them. But by recognizing they are not aligned with who you are now. And choosing not to engage.

The New Identity Becomes Normal

As you continue to remain with the new identity, something shifts. It becomes normal. Not something you're maintaining. Not something you're thinking about. Just something you are.

And once it feels normal, it stabilizes. At that point, it no longer feels like a shift. It feels like reality.

You Don't Need to Protect It

Another misconception is that you need to "protect" your new identity.

That you need to: avoid certain thoughts, stay perfectly aligned, constantly reinforce it.

But identity doesn't need protection. It needs acceptance. Once accepted, it sustains itself.

What About Doubt?

Doubt can appear. That's normal. But doubt only matters if you identify with it.

If you see doubt and think:

"This means I'm not there yet," you reconnect with the old identity.

If you see doubt and recognize:

"This is just a leftover thought," it passes.

Without effect.

The Simplicity of Stabilization

At its core, stabilization is simple:

Continue identifying as the version you selected. Not perfectly. Not forcefully. Just consistently.

You Are Not Maintaining—You Are Being

This is the key shift. You are not maintaining a new identity. You are being it.

Maintenance feels like effort. Being feels natural. And the more you allow yourself to be that version, the more stable it becomes.

The Role of Time

Time does not stabilize identity. Consistency of identification does.

Time only appears to matter because people move back and forth between identities. But when identity is consistent, stability is immediate.

What This Means for You

You don't need to worry about losing your new reality. You don't need to constantly check yourself. You don't need to analyze every thought.

You only need to recognize:

Who am I identifying as right now?

And if that answer remains clear— your reality remains stable. Because stability is not something you achieve. It is something that happens when identity no longer shifts back.

The Simplicity of It

You stabilize a new reality by refusing to return to the old identity. Not through effort. Not through control. But through clarity.

And once that clarity is steady— your reality reflects it just as steadily. Because it always has. Now, you're simply choosing it on purpose.

WHO AM I BEING

Chapter 12

The Simplicity of Self-Concept

By now, you've seen this idea from multiple angles. You've seen how identity reflects through reality. How thoughts follow identity. How circumstances are not authority. How one decision can shift everything.

And at this point, it can almost feel like there are a lot of pieces. A lot to understand. A lot to remember. A lot to apply.

But here's what all of it comes down to:

Self-concept is simple.

Not because there is nothing to it. But because everything you've read points back to the same core truth.

One Idea Behind Everything

No matter what area of life you look at—
money
relationships

health
success
confidence
stability

The same principle is operating:

Your reality reflects who you believe you are.

Not in parts. Not selectively. But completely.

Why It Seems Complicated

The reason this can feel complicated is because of how many ways it shows up.

You might experience:
different challenges in different areas
different patterns in different situations
different emotions depending on the context

And it can feel like each one needs its own solution. But they don't. They all trace back to identity.

One Identity, Many Reflections

You don't have separate identities for every part of your life. You may express identity differently in different areas. But at the core, it is still one self-concept. And that self-concept reflects across everything.

For example:

If you see yourself as someone who has to work hard for results, that may show up in:
money
relationships
personal goals

If you see yourself as someone things always work out for, that will also show up everywhere.

Not just in one category. Because identity is not compartmentalized. It is consistent.

You Don't Need Separate Solutions

This is where simplicity returns.

You don't need:
one method for money
another for relationships
another for confidence

You don't need to solve each area individually. Because each area is reflecting the same identity. So when identity changes, everything shifts.

The Power of General Identity

This is also why broad identity statements are so effective.

Statements like:

"I am someone things always work out for."

"My life works in my favor."

"Everything aligns for me."

These don't target one area. They define the whole identity. And because of that, they influence everything.

Why Specific Fixes Don't Last

When people try to fix specific problems, they often see temporary change. But the underlying identity remains the same.

So eventually, something else appears that matches that identity. It may look different. But it feels the same. Because the root hasn't changed.

Returning to the Core

Instead of trying to solve every situation, you return to the core question:

Who am I?

Not in a philosophical sense. But in a practical sense.

Who am I being in my life?

Because that answer determines everything that follows.

The Simplicity Is the Power

It can feel like this idea is too simple. That there should be more to it. That something is missing. But simplicity is what makes it powerful.

Because it removes:
confusion
overthinking
unnecessary effort

You don't need to track every detail. You don't need to manage every thought. You don't need to analyze every situation. You only need to recognize identity.

You Already Know How This Works

This is not something you are learning for the first time. You have been living this.

You have already seen:
patterns repeat
certain things feel normal
certain outcomes feel expected

All of that has been guided by self-concept. Now you are simply becoming aware of it.

Applying It Everywhere

Once you understand this, you can apply it anywhere.

If something shows up in your life, you don't need to ask:

"How do I fix this?"

You ask:

"What identity is this reflecting?"

And:

"Is this the identity I want to continue with?"

If the answer is no, you choose differently. And that choice affects everything.

There Is Nothing Separate

There are no separate problems. There are no isolated issues. There is only identity expressing itself in different ways. Once identity shifts, those expressions shift with it.

The Illusion of Complexity

The mind often wants to make things complex. It wants steps. It wants methods. It wants detailed instructions.

But complexity comes from trying to control outcomes. Simplicity comes from understanding the source.

Living From Simplicity

When you begin to live from this understanding, something changes. You stop chasing solutions. You stop trying to fix every detail. You stop reacting to everything that appears.

And instead, you remain anchored in identity.

Because you know:

That is what determines everything.

The Simplicity of It

At its core, this is all there is:

Your reality reflects your self-concept.

Not sometimes. Not in certain areas. But always.

What This Means for You

You don't need to manage your life piece by piece. You don't need to solve every problem individually. You don't need to overthink the process.

You only need to recognize: Who you are being.

And once that is clear—everything else follows. Not because you made it happen. But because self-concept applies to everything. And once you understand that—everything becomes simple again.

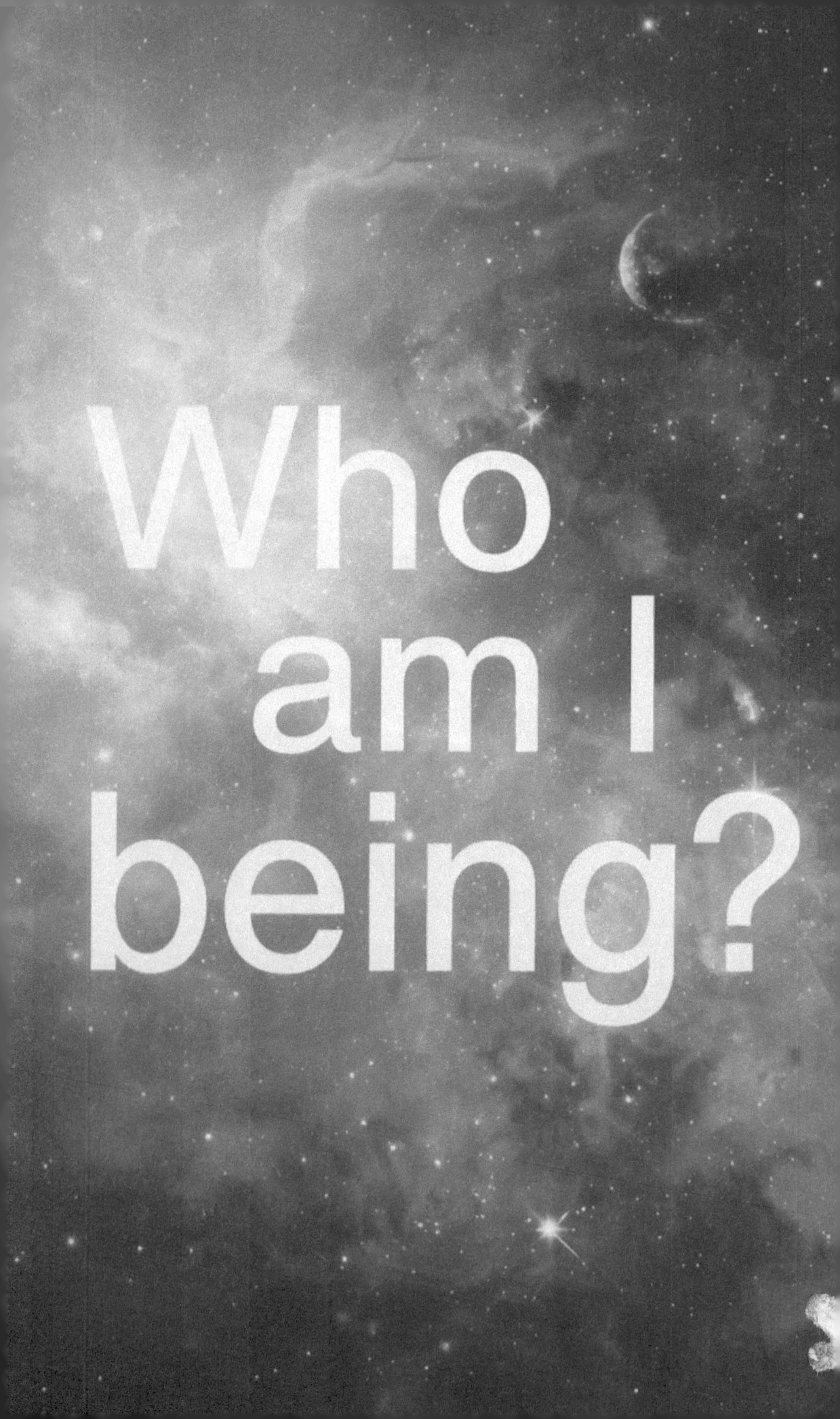
Who am I being?

Final Chapter: Self-Concept Selects Your Reality

At the beginning of this book, you were introduced to an idea:

That your reality is not something you are trying to change piece by piece. It is something that is reflecting who you are.

Now, at the end, that idea is no longer just a concept. It is something you can see. Something you can recognize. Something you can use.

Everything Comes Back to This

No matter what you've read in this book, every chapter has been pointing to the same place. ...rent techniques. Not to different

...nderstanding: Self-concept se... ...lity.

Not creates it over time. Not builds it step by step. Selects it.

The Position You've Been In

Before this understanding, it likely felt like you were:
trying to change your life
trying to fix situations
trying to manage thoughts
trying to control outcomes

Even if you understood some of these ideas, there was still a sense of:

"I need to do something to make this happen."

That position places you inside the experience. Reacting to it. Responding to it. Trying to adjust it.

The Position You Are In Now

Now, the position shifts. You are no longer inside the experience trying to control it. You are at the level where it is selected.

You are not reacting to reality. You are recognizing what is reflecting—and choosing differently. This is a completely different plac operate from.

Because it removes:
struggle
confusion
the need to fix everything

And replaces it with clarity.

You Are the Selector

You are not waiting for something to happen. You are not hoping for change. You are not trying to become something over time.

You are selecting who you are. And that selection determines everything else.

There Is No Separation

There is no separation between:
identity and experience
self-concept and reality
who you are and what you live

They are the same thing. Seen from different angles. Once identity is accepted, experience follows.

The Simplicity of Selection

Selection is not complicated. It is not something you need to learn how to do. You are already doing it.

Through:
what you believe
what you assume
what you declare
what you accept as true

The difference now is awareness. You are no longer selecting unconsciously. You are selecting intentionally.

You Don't Need to Manage Reality

You don't need to:
control outcomes
fix circumstances
monitor every thought
analyze every situation

Because those are all reflections. And reflections don't need to be managed. They change when the source changes.

The Source Is Always Identity

Everything you experience can be traced back to identity. Not in a complicated way. But in a direct way.

If something appears, it is reflecting something about who you are assuming yourself to be.

And that assumption can change. At any moment.

You Are Free From the Old

You are not bound to:
your past
your patterns
your previous experiences
what has already happened

Because none of those things are selecting your reality. They are reflections of what has already been selected. And selection can change instantly.

This Is Not About Effort

This is not about doing more. This is not about trying harder. This is not about perfecting a process.

This is about recognizing what has always been happening—and choosing it consciously.

Living From This Understanding

When you live from this understanding, something shifts.

You stop chasing outcomes. You stop reacting to circumstances. You stop questioning every detail.

And instead, you remain clear on one thing:

Who am I?

Because that answer determines everything.

There Is Nothing Else to Manage

Once this is clear, there is nothing else to manage. No separate areas. No separate problems. No separate solutions.

Just identity. And the reality that reflects it.

The Simplicity of It

At its core, this is all there is:

Self-concept selects your reality. Not partially. Not occasionally. But completely.

What This Means for You

You don't need to go back and relearn everything. You don't need to find new methods. You don't need to search for more answers. You already have what matters. You understand the source.

And once the source is understood, everything becomes simpler.

Final Position

You are not someone trying to change your life. You are someone who selects it. And from that position—everything becomes clear. Because nothing is happening to you.

Everything is reflecting you.

And now—you are choosing that reflection on purpose.

About The Author

The Quantum Shifter is the pen name of an Ivy-trained psychologist, researcher, and lecturer whose work explores the intersection of identity, consciousness, and reality creation. After completing advanced studies in psychology and spending years teaching and writing at the university level, she shifted her focus toward the deeper question that shaped her life's work:

How does identity create reality — and why does changing the story of who you are change the world you live in?

Her publications draw on decades of private research, academic inquiry, and experiential observation. Known for her clarity, precision, and deeply grounded approach, she writes about identity physics, timeline selection, self-concept, and the mechanics of reality in a way that is both accessible and transformative.

Although her academic background is traditional, her pen name reflects the evolution of her work — a nod to the moment she realized that consciousness itself is the most powerful instrument of change. The name stuck, and readers around the world now associate *The Quantum Shifter* with high-level insight, elegant simplicity, and rapid identity transformation.

She currently lives a quiet, contemplative life dedicated to writing, research, and teaching the principles of identity-based reality creation. Her books serve as distilled explorations of the concepts she has spent decades studying — written for readers who are ready to shift who they are, not just what they want.

www.ingramcontent.com/pod-product-compliance
Lightning Source LLC
Chambersburg PA
CBHW031327230926
51848CB00001B/122
9781972432198